DIAMOND GRINDING

Diamond Grinding—Recent Research and Developments deals with three interesting examples of work in this area and their effects on grinding efficiency—metal cladding abrasives, the use of different coolants, and machine stability and vibration effects.

Diamond grinding wheel users will not only find this book of interest, but may also derive considerable economic benefits from some of the results obtained.

All the research work described has been carried out at the De Beers Diamond Research Laboratory in Johannesburg, where all aspects of diamond research, from exploration, mining and recovery to gem cutting and industrial diamond utilization are undertaken.

Other titles in this series

ME/1 TRIBOLOGY
E D Hondros

ME/2 ELECTRON BEAM WELDING
M J Fletcher

ME/3 VACUUM BRAZING
M J Fletcher

ME/4 STRESS CORROSION FAILURE
P Greenfield

ME/5 GLOW DISCHARGE MATERIAL PROCESSING
R A Dugdale

Other Mechanical Engineering titles available

TL/ME/1 HYDROSTATIC EXTRUSION
J M Alexander and B Lengyel

TL/ME/2 THE HEAT PIPE
D Chisholm

All published by Mills & Boon Limited

M & B Monograph ME/6

General Editor: J. Gordon Cook, PhD, FRIC

Diamond Grinding

Recent Research and Developments

J Burls, BSc

Director, Industrial Diamond Information Bureau

Mills & Boon Limited

London

First published in Great Britain 1971
by Mills & Boon Limited, 17–19 Foley Street,
London, W1A 1DR

ISBN 0.263.51791.8

Made and printed in Great Britain
by Butler & Tanner Limited, Frome and London

CONTENTS

1. Introduction

The emergence of grinding in this century as a precision production-line technique, rather than as an ancillary to production, has been one of the factors which have helped make possible the reproducibility of parts essential to mass production.

One type of grinding wheel of particular importance, the diamond grinding wheel, first came on the scene some 50 years ago, following the introduction by industry of the ultra-hard cemented carbides. Beginning as a toolroom adjunct devised to manufacture and shape tungsten carbide tools, it has now achieved an importance to industry which is impossible to assess. It is used on a bewildering range of materials, both hard and soft, including stone, glass and ceramics; it is to be found in virtually every branch of the engineering and allied industries. Diamond grinding is today the most important and widely used of all diamond machining processes.

In the last 30 years, many new ultra-hard and abrasive materials have been introduced which are extremely difficult to shape and machine economically using conventional abrasives. As these new materials have come along, so have the diamond tool makers made steady progress with improvements in materials and methods to produce better and more efficient wheels to cope with them. Paralleling this development has been the production of vastly improved grades of diamond abrasives, especially since the advent of diamond synthesis. Other research work into operating conditions, particularly at the Diamond Research Laboratory in Johannesburg, has helped this progress.

Although improvements in diamond grinding wheel performance have been continual, especially since the 1940s, there have been some lines of research that have led or will lead to rapid improvements in efficiency. Three of these—the introduction of metal-clad diamond, grinding machine stability, and coolants—are discussed in this monograph.

2. The Metal-Cladding of Diamond Abrasives

In the early days of diamond grinding, the diamond in the wheels was graded only by size. Wheelmakers commonly purchased crushed natural diamond in parcels containing a variety of sizes, and subsequently sorted for size only. The first step towards a more sophisticated sub-division was taken during the mid-1950s, when wheelmakers began to use the more blocky particles of diamond in metal bonds and the flatter, more needle-shaped particles in the weaker resin bonds. In general terms, the former gave longer wheel life and the latter faster grinding speeds.

The advent of diamond synthesis in 1953 led to an increase in the range of diamond abrasives available to industry. An immense development programme by producers of both natural and synthetic diamond resulted in considerable improvements in diamond quality and utilization.

A dramatic development, in the mid-1960s, was the introduction of metal-clad diamond for resinoid grinding. Work on this project by the De Beers research groups led in 1966 to the publication of results which were both spectacular and technically extremely interesting. Briefly, they disclosed that the effect of covering individual particles of diamond grit with a coating of metal (55% by weight of the coated diamond particle) was to increase, when wet grinding, the efficiency of resinoid diamond grinding wheels by several hundred per cent. Significantly, this marked increase in grinding efficiency was achieved with little loss of those characteristics which are of major importance to resinoid diamond grinding—namely, that the wheel should be cool and free-cutting and impart a good surface finish.

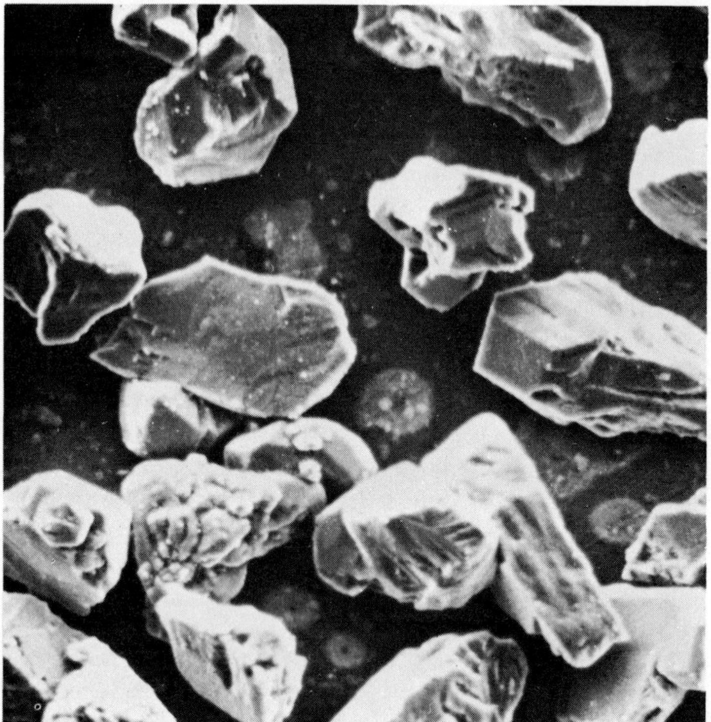

Fig. 1. De Beers RDA 120/140 US mesh. Mag. 100×. A synthetic diamond abrasive containing irregular shapes with rough surfaces, engineered for resin bond grinding wheels used mainly to machine hard metal carbides.

Fig. 2. De Beers RDA55N 100/120 US mesh Mag. 140× (top). A synthetic diamond abrasive containing irregular-shaped particles clad with a special metal coating. The optical micrograph of a resin bond wheel containing RDA55N (bottom, Mag. 450×) illustrates the excellent bond retention of the clad grit.

The wet surface grinding of tungsten carbide and combinations of tungsten carbide and steel with metal-clad grits is now carried out daily in industry the world over. The average efficiency, and therefore the life, of wheels containing metal-clad grits has increased by roughly 3 to 1 compared with wheels containing the best of the unclad grits. The cost of wet surface grinding has therefore been reduced considerably. (Alternatively, of course, by increasing feeds and speeds the machine output can be increased, with no increase in wheel costs and with the added bonus of a reduction in time costs.)

WET GRINDING EFFICIENCY TESTS

The results of one series of tests carried out at the Diamond Research Laboratory in Johannesburg illustrate effectively the improvements in wet grinding efficiency brought about by using metal-clad diamond. Test parameters were:

Machines	3 Jones and Shipman Model 540 tool room surface grinders with 2 hp spindle drive motors.
Wheel type	D1A1
Wheel size	5 in × $\frac{1}{4}$ in × $1\frac{1}{4}$ in
Wheel speed	4,000 rev/min
Wheel peripheral speed	5,200 ft/min (26·4 m/sec)
Diamond types	De Beers RDA and RDA55N
Diamond mesh size	60/80 to 325/400
Diamond concentration	100 (72 cts per cubic inch)
Bond depth	$\frac{1}{8}$ in (3·1 mm)
Bond hardness grade	R
Downfeed	0·001 in per pass (0·025 mm)
Total downfeed per test	0·120 in (3·0 mm)
Crossfeed	0·060 in per table reversal (1·5 mm)
Table traverse speed	53 ft/min (16 m/min)
Coolant	Water + Bryto 5 (100 : 1)

Coolant flow rate	40 gallons per hour (3 litres/min)
Sparkout	2 sparkout passes, without downfeed, at the end of each test run.

TUNGSTEN CARBIDE SPECIMENS

These were made up of 18 pieces, each 2 in × 1 in × ½ in (50·8 × 25·4 × 12·7 mm), arranged as a block with a face area of 6 in × 3 in (18 sq in) (152·4 × 71·2 mm) (116 sq cm) and a height of 1 in (25·4 mm). The ISO group classification of the grade of carbide used was K40.

TEST CONDITIONS

Because the surface condition of a wheel can have a considerable effect upon test results, all the wheels were given a conditioning run on the material to be ground and at the machine parameters to be used before the test proper was started. At the end of each test a sparkout, without downfeed, of two full passes across the specimen was given.

After each conditioning or test run was completed, the wheels were returned to a measuring room where they remained for at least 24 hours before measurements were taken. This procedure enabled the wheels to stabilize at a temperature of 68°F ± 1°F (20°C ± ½°C) and thus overcome any deleterious effects caused by thermal expansion. The rest period also allowed the wheels to overcome the effects of centrifugal expansion.

The amount worn from a wheel during a test was measured in a special jig and the results were reproducible to ±0·000,020 in (0·5 microns). The minimum amount removed from the wheels during any particular test was 0·004 in (0·1 mm). Measurement was made by dial depth indicators at each corner of the ground block before and after each test. The average depth ground for one test was 0·120 in (3·0 mm). The wheels were run in pairs and at least three tests were done at any single grinding condition. Each figure quoted is, therefore, the average of at least six tests.

The wheel efficiency or *G* ratio was calculated from the volume of material ground away per unit volume of bond worn from the wheel during testing.

$$G = \frac{\text{Volume material removed (in}^3\text{) (cm}^3\text{)}}{\text{Volume wheel worn away (in}^3\text{) (cm}^3\text{)}}$$

The coolant flow rate was held at 40 gallons per hour (3 litres per minute) throughout all tests, measured by rotameters. Coolant tanks were replenished with pre-mixed fluid at the start of each shift and completely renewed after 100 hours of grinding. Each machine was equipped with a centrifuge to clarify the used coolant; the centrifuge bowl, containing the grinding sludge, was cleaned at the end of each test.

The total power used during each test was measured by kilowatt hour meters and a continuous record of power drawn was kept using recording wattmeters. In order to remove discrepancies due to changes in test times, the figures quoted are for one hour's work at the particular parameters in use (kWh).

Wheel cost, in the case of all the 100 concentration wheels, has been estimated at a rate of £170 ($410) per cubic inch of diamond-bearing bond. Time cost, which includes machine, operator and overhead costs, has been estimated at the rate of £3 ($12.00)* per hour.

The surface finish obtained on the ground blocks was measured with Hommel-Tester Type T equipment. The figures quoted are AA or middle roughness value (CLA. Ra).

Table 1 and Fig. 3 show the results of the tests. They demonstrate clearly the increase in efficiency of the metal-clad over the unclad grits. In these tests, the wheels were run at a downfeed of 0·001 in (0·025 mm). As a wheel containing the smaller grits would not normally be punished by using such a large downfeed, tests were included at a downfeed of 0·0004 in (0·01 mm) for 170/200

* £3 and $12.00 per hour are average labour costs for the U.K. and U.S.A. respectively.

Table 1

GRIT SIZE	*G* RATIO		PERCENTAGE INCREASE IN *G*	POWER DRAWN Kwh	
	UNCLAD	CLAD		UNCLAD	CLAD
			0·001 in downfeed		
60/80	158	267	69	0·64	0·85
80/100	134	255	90	0·63	0·83
100/120	109	244	124	0·61	0·80
120/140	100	230	130	0·60	0·77
140/170	78	207	166	0·59	0·73
170/200	59	160	172	0·57	0·68
200/230	40	124	210	0·55	0·63
230/270	32	110	244	0·54	0·60
270/325	27	98	263	0·53	0·58
325/400	21	74	352	0·52	0·54
			0·0004 in downfeed		
170/200	120	212	76	0·57	0·63
200/230	94	168	79	0·55	0·59
230/270	80	156	95	0·54	0·58
270/325	66	136	106	0·52	0·56
325/400	52	116	123	0·50	0·54

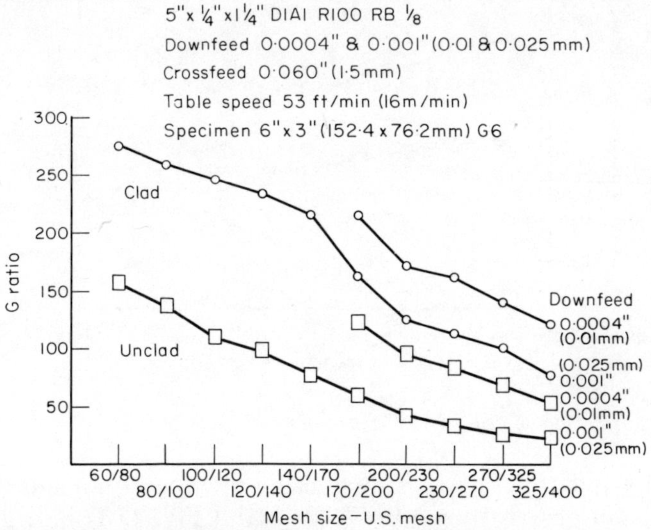

Fig. 3. Wheel efficiency—clad and unclad grits.

mesh grits and smaller. The figures from these tests are included in Table 1 and shown graphically in Fig. 3.

An interesting point emerges from the tests conducted at the 0·001 in (0·025 mm) downfeed. The highest efficiency achieved with the unclad grits, 158, was with 60/80 mesh. Comparable efficiency with the clad grit was obtained when using 170/200 mesh grit. It is necessary to consider carefully, therefore, the grit size to be used for a particular job. Prior to the advent of metal cladding, one might have used, say, a wheel containing 100/120 mesh grit to do a rough removal job and then changed to a 325/400 mesh grit to take off the last 0·0004 in (0·01 mm) and to generate the required surface finish; it is now possible, however, to take a wheel containing 200/230 mesh clad grit to do both the rough removal and the finishing. The use of the finer metal-clad grits for the two jobs of rough removal and surface finishing is obviously very advantageous for some wheel users.

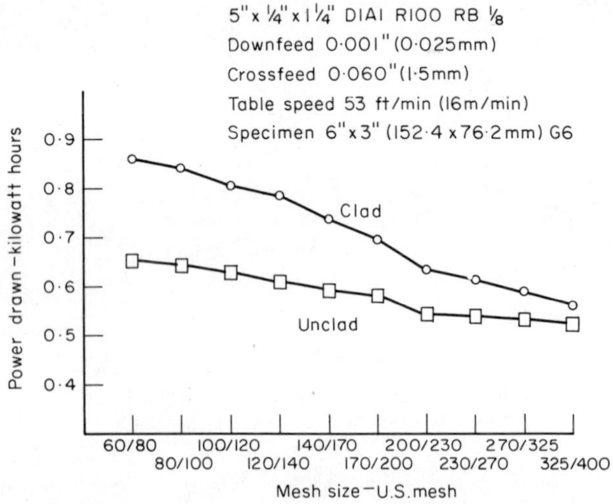

Fig. 4. Power drawn—clad and unclad grits.

The amount of power drawn using the metal-clad and the unclad grits is shown in Table 1 and Fig. 4. The metal-clad grits use more power than the unclad grits, the differences ranging from 33% more in the case of the 60/80 mesh size, down to 4% more in the 325/400 mesh size. This higher power demand by the metal-clad wheels is partly due to the fact that the resin bond matrix can adhere more effectively to the metal coat around the grit than to the diamond grit itself; thus, more grit particles are held in the bond, and kept working, than is the case with the unclad grits. This has a disadvantage: where a machine motor is just able to cope with unclad grit wheels, the extra power required by the metal-clad grits could cause the motor to stall. It should be noted that the amount of power drawn, as shown in Table 1, is the total power used; it therefore includes "idling power", at the rate of 0·24 kWh.

The efficiency, or *G* ratio, figures obtained when using a diamond abrasive grinding wheel do not always give the full picture of a grinding operation. The *G* ratio figure on its own can be used only when the wheels being compared are exactly the same size and cost, and identical machine parameters are used for all the tests in the evaluation. Under these conditions the wheel cost is of prime importance. Time cost changes the picture by a much smaller factor, as is evident from the figures shown in Tables 2 and 3. As soon as machine parameters are changed, the time cost picture becomes of greater importance.

COST ANALYSIS

The cost analysis of the tests is shown in Tables 2 and 3, and in Fig. 5. Tables 2 and 3 show that, for the removal of a particular amount of tungsten carbide, the wheel cost has a considerable bearing on the total cost. If the wheel efficiency drops by 50% then the wheel cost component of total cost increases by 100%, whereas the time taken to remove the material increases by only a very small amount.

For example, for a *G* ratio of 158, with a 60/80 unclad grit wheel, the wheel cost is £1·08 per cubic inch of material removed, but a 140/170 unclad grit wheel with a *G* ratio of 78 (a drop of 50%) has a wheel cost of £2·18, an increase of 100%. The time costs, however, are £3·37 and

Table 2. Unclad Grits

GRIT SIZE	*G* RATIO	REMOVAL PER IN³ TIME (mins)	TIME COST	* WHEEL COST	TOTAL COST
		0·001 in downfeed			
60/80	158	67·3	£3·37	£1·08	£4·45
80/100	134	67·6	£3·38	£1·27	£4·65
100/120	109	68·1	£3·41	£1·56	£4·97
120/140	100	68·4	£3·42	£1·70	£5·12
140/170	78	69·2	£3·46	£2·18	£5·64
170/200	59	70·3	£3·52	£2·88	£6·40
200/230	40	72·8	£3·64	£4·25	£7·89
230/270	32	74·8	£3·74	£5·32	£9·06
270/325	27	76·5	£3·83	£6·30	£10·13
325/400	21	79·6	£3·98	£8·10	£12·08
		0·0004 in downfeed			
170/200	120	169·6	£8·48	£1·42	£9·90
200/230	94	171·3	£8·57	£1·81	£10·38
230/270	80	172·7	£8·64	£2·15	£10·79
270/325	66	174·7	£8·74	£2·58	£11·32
325/400	52	177·8	£8·89	£3·27	£12·16

* All costs are given in new decimal currency.

£3·46 respectively, giving the 140/170 unclad grit wheel an increase of only 2½% over the 60/80 unclad grit wheel.

On the same basis, the 200/230 clad grit wheel, with a *G* ratio of 124, gives a total cost of £4·76 per cubic inch of material removed, against the 100/120 clad grit total cost of £4·03, *G* ratio 244. This again represents a 100% wheel cost increase (£0·70 to £1·37) for a *G* ratio drop of 50%. Again, the difference in time cost, from £3·33 to £3·39, gives an increase of only 2% when using the 200/230 in place of the 100/120 clad grit wheel.

A point of considerable interest concerns the use of smaller downfeeds when using wheels containing fine grits. Table 3

Table 3. Clad Grits

GRIT SIZE	G RATIO	REMOVAL PER IN^3			
		TIME (mins)	TIME COST	* WHEEL COST	TOTAL COST
		0·001 in downfeed			
60/80	267	66·5	£3·32	£0·64	£3·96
80/100	255	66·6	£3·33	£0·67	£4·00
100/120	244	66·6	£3·33	£0·70	£4·03
120/140	230	66·7	£3·34	£0·74	£4·08
140/170	207	66·8	£3·34	£0·82	£4·16
170/200	160	67·3	£3·37	£1·06	£4·43
200/230	124	67·8	£3·39	£1·37	£4·76
230/270	110	68·0	£3·40	£1·55	£4·95
270/325	98	68·5	£3·42	£1·73	£5·15
325/400	74	69·4	£3·47	£2·30	£5·77
		0·0004 in downfeed			
170/200	212	166·9	£8·35	£0·80	£9·15
200/230	168	167·8	£8·39	£1·01	£9·40
230/270	156	168·2	£8·41	£1·09	£9·50
270/325	136	168·9	£8·45	£1·25	£9·70
325/400	116	169·8	£8·50	£1·47	£9·97

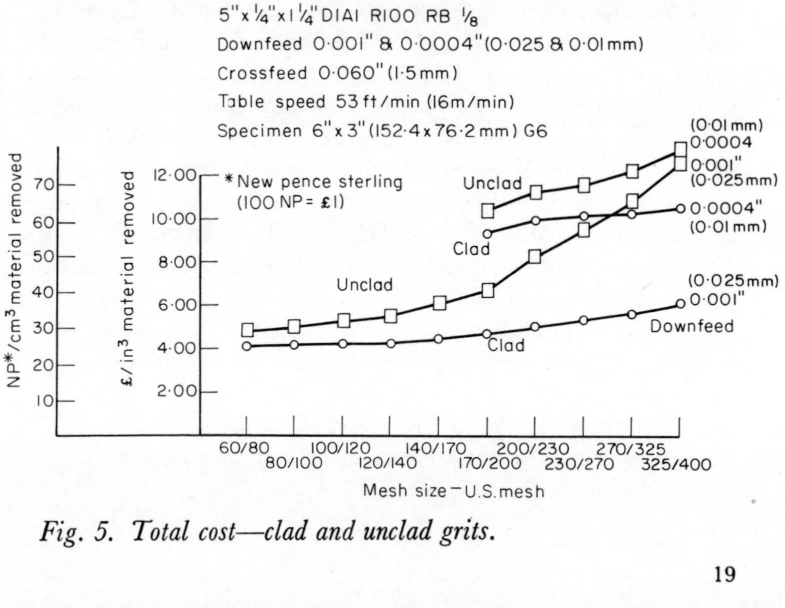

Fig. 5. Total cost—clad and unclad grits.

shows that decreasing the downfeed from 0·001 in to 0·0004 in, when using for example 325/400 mesh grits—the smallest in the range—increases the cost of removing one cubic inch of material from £5·77 to £9·97—an increase of 73%. This exposes a very glaring malpractice—"babying" the wheel, to make it last longer.

The total costs of the same grit sizes clad and unclad are of particular importance. Examples are shown in Table 4, for 0·001 in downfeed. Obviously, the use of metal-clad grits should *always* be advocated for a wet surface grinding operation.

Table 4

GRIT SIZE	TOTAL COST (£/in³) MATERIAL REMOVED 0·001 in downfeed	
	UNCLAD	CLAD
60/80	£4·45	£3·96
325/400	£12·08	£5·77

One of the more interesting new metal-clad diamond abrasives is De Beers DXDA-MC. With this grit, diamond grinding of cast iron, and ductile materials such as steel, is now a commercial proposition. As an example, DXDA-MC diamond grinding wheels have proved superior to conventional methods for grinding Udimet 500, a nickel-base alloy containing chromium, cobalt, titanium, molybdenum and other metals. With a DXDA-MC wheel Mercury Machines of Cleveland, Ohio, achieved eight times more stock removal on Udimet 500 at one pass than they had previously obtained with an aluminium oxide wheel; this resulted in a 30% saving in time and labour costs. Only 14 parts could be ground with the aluminium oxide wheel between dressings. With DXDA-MC wheels, 900 to 1,000 parts were ground before redressing became necessary.

Fig. 6. De Beers DXDA-MC 80/100 US mesh, Mag. 100×. A metal-clad diamond abrasive, specially developed for grinding cast iron, steel and ductile metals. Its strong blocky particles present cutting edges at the correct angle to workpiece, and operate by chip-removal rather than by abrasion. The metal-cladding promotes retention in the resin bonds for which this grit was designed.

At Fluid Regulators, another American company, stainless steel sleeves for Lockheed's jet engine thrust reversers are made, using DXDA-MC, in exactly half the time and with five times the stock removal rate of aluminium oxide previously used. Labour cost has been reduced from \$10.00 to \$5.00 per sleeve.

Myford Limited, of Nottingham, UK, cut the cost of finishing cast iron lathe beds by 40% by using a grinding

Fig. 7. A 10-in diameter resin bond wheel containing 80/100 US mesh De Beers DXDA-MC solved a long-standing problem at Metal Box Co., by its ability to grind extremely hard SC 26 and ductile NRM steel with equal ease. Used on a Jones & Shipman 1310 EIUR surface grinder with a copious supply of Hocut soluble oil coolant, the wheel easily maintained the 0·0001 in on diameter required on punch tools in both these steels.

wheel impregnated with DXDA-MC. Previously Myfords produced a surface finish acceptable visually and in flatness on the bed of their Loughborough Training Lathe in two operations—17 minutes of grinding with a silicon carbide wheel (purchase price £7·53) followed by 45 minutes of scraping. Now, using a 10 in. diameter DXDA-MC flared cup wheel (cost £64) grinding time is reduced to 9 minutes, and no further work is required, since the DXDA-MC wheel alone produces a visually acceptable 20 micro-inch finish. The cost of the operation is reduced from £4·16 to £2·50, in spite of the difference in wheel costs, by a reduction of production time from over an hour to 9 minutes.

RDAR50C

RDAR50C, the latest metal-clad grit to come on the market, was developed specially for dry grinding conditions. It involved the synthesizing of an essentially new quality of diamond which, after processing, is covered with a composite metal coating developed by De Beers scientists. Comprehensive tests at the Laboratory showed that the ideal particle for dry grinding should be less strong—more friable—than that developed for wet grinding. A special manufacturing process was developed to make a product with the desired friability. Simultaneously, the Laboratory embarked on a full research programme to determine the metal cladding most suited to dry grinding. It was soon found that no single metal coating was entirely suitable, and a composite cladding of several layers of different metals or alloys, each selected to confer a specific desirable property to the clad grit, was evolved.

The end product of this research, RDAR50C, has achieved the highest dry-grinding performance yet in laboratory and field evaluation. Perhaps the most convincing of many tests was with ISO group P20 tungsten carbide. This grade of carbide is one of the most difficult of all to grind dry because of the chipping and breakage associated with the high TiC–TaC concentration. A series of 5 in by $\frac{1}{8}$ in

Fig. 8. De Beers RDAR50C 80/100 U.S. mesh, Mag. 100×. Resinbond Diamond Abrasive (RDA) of reduced strength (R) coated to 50% with a special composite metal cladding (50C). This new material is specially designed for the dry grinding of metal carbides, an operation in which it outperforms all other diamond abrasives available up to the present time.

face D11V9 wheels was made up at 100 concentration (72 carats per cubic inch or 4·4 carats per cubic centimetre of wheel bond) in each of the mesh sizes from 60/80 to 325/400. In the tests, an indexing head was used to simulate the regrinding of milling cutter inserts. Severe grinding conditions, an infeed of 0·002 in and a table speed of 6 ft per minute, were deliberately chosen.

Full details of the tests and the results obtained are shown in Figure 10. To provide a basis for comparison, the results of a second series of tests are also shown. A feature of these curves is that there is a broad performance optimum generally lying between 100 and 230 mesh. The

Fig. 9. Using a wheel containing De Beers RDAR50C metal-clad diamond abrasive for dry grinding tungsten carbide tipped milling cutters, a London factory found that wheel life was increased by approximately one-third.

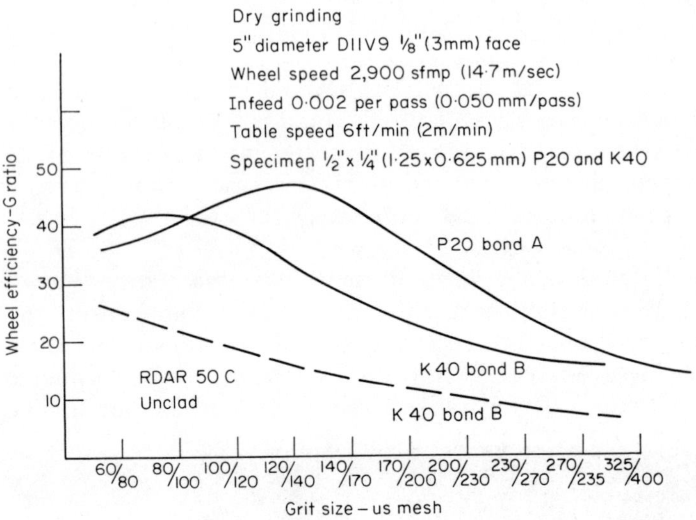

Fig. 10. The relative efficiency of RDAR50C in two different bonds, and an unclad grit while grinding two grades of carbide dry. P20 carbide is prone to spalling at 0·002 in (0·05 mm) infeed with any grit other than RDAR50C.

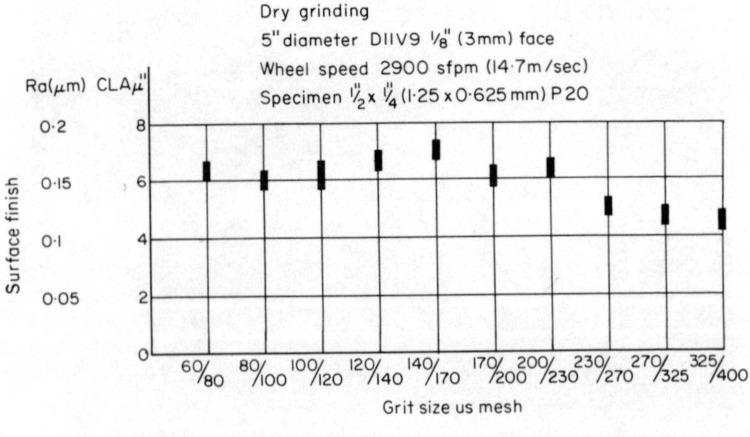

Fig. 11. Surface finish achieved on P20 carbide inserts set in a simulated milling cutter. Finish grinding was carried out manually under normal tool-shop grinding conditions.

position of this optimum changes with changes in grinding conditions, diamond concentration, workpiece grade, etc.

The surface finish measured for each of the full range of grit sizes is shown in Fig. 11. RDAR50C gives a satisfactory finish and, as may be expected, the finer grits produce finer finishes.

Power drawn is a most important factor in the use of diamond wheels. Relative power values measured during these tests are shown in Fig. 12. It is interesting to note how closely the power drawn follows the *G* ratio curve. This is, of course, generally true of diamond grinding—the higher the *G* ratio, the greater the power drawn. However, while RDAR50C requires higher power than unclad grits, the grinding action of the wheels containing RDAR50C is exceptionally cool and free cutting. Even when grinding the difficult P20 grade, there is no sign of the burning, chipping or breakage usually associated with the dry grinding of this carbide type—undesirable results which can normally only be avoided by using greatly reduced volumetric removal rates.

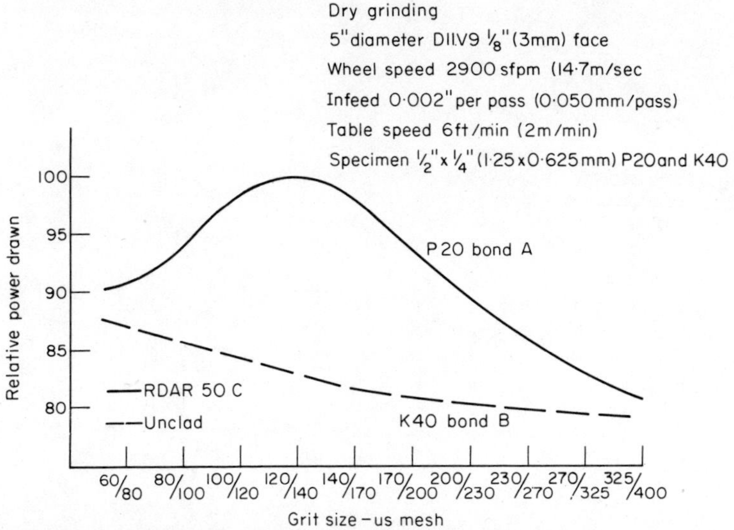

Fig. 12. Relative power values measured during the tests for clad and unclad grits.

HOW DOES THE COATING WORK?

De Beers' extensive research into the fundamentals of grinding has indicated three main advantages of cladding selected diamond grits with a coating of metal:

1. The metal cladding acts as a heat sink to dissipate the heat generated at the grinding point, thus minimizing damage to the resin.
2. It forms a strong jacket around each particle, as a result of which, even after particle fracture, the remaining diamond fragments are retained in the wheel to continue doing useful work.
3. It improves bond retention since metal–resin adhesion is generally superior to diamond–resin adhesion.

3. Machine Stability and Vibration Effects

As with most other tools, there are certain optimum conditions under which the diamond abrasive grinding wheel will operate with maximum efficiency. Ideally, diamond wheels would be used on machines designed especially for them. In practice, they are usually used on the machines that are available. In many cases, these machines lack the rigidity and stability necessary for efficient diamond wheel usage.

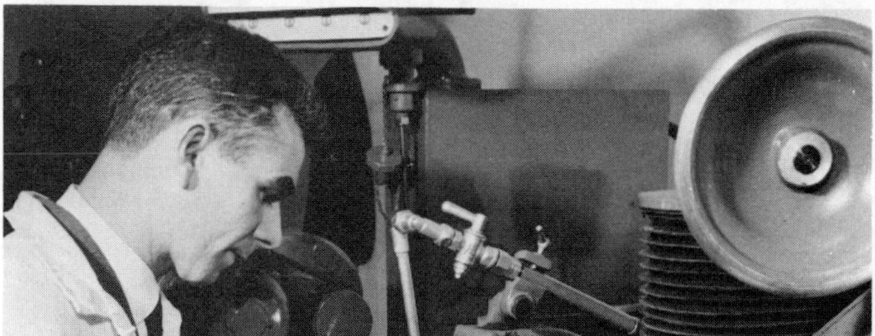

Fig. 13. The Jones & Shipman 1310 EIUR with its rigid plain-bearing wheel-head is a very popular universal grinding machine adaptable for both internal and external grinding. It can also be used if the occasion arises for tool and cutter grinding.

Fig. 14. The Jones & Shipman 1411 SF surface grinding machine is of extremely rigid construction and has sufficient horse-power to utilize the free-cutting characteristics of diamond abrasive grinding wheels.

Some of the adverse effects of machine instability are already widely known, and they display themselves daily in machine shops and tool rooms all over the world. The true importance of machine stability for diamond wheel users, however, was made apparent with the publication of the results of a series of tests by the De Beers Diamond Research Laboratory. They proved conclusively that much higher levels of efficiency and corresponding reductions in costs are obtained with a stable machine.

It was known to the Diamond Research Laboratory that loss of efficiency when grinding with diamond abrasive wheels could often be attributed to machine defects, and not to the wheel itself. In addition to a lack of machine stability, these defects included vibration pick-up from the drive, hydraulic motors and belts, worn or insufficient spindle drive bearings, and stickiness or shuddering in the table traverse movement.

The test programme was designed to compare results obtained on a machine which had these defects with results obtained on a defect-free machine, and to evaluate the results in scientific terms. To do this, test grinding was carried out on two machines which, although of different makes, were both standard models carefully selected to produce the wide mechanical differences required.

Machine A, which had been in use in the Laboratory for about two years, was chosen to represent a standard cutter grinder in a condition typical of that to be found in industry today. This machine, a surface grinder fitted with an automatic infeed and indexing system, had from the grinding efficiency point of view certain undesirable characteristics. In the words of the DRL report:

> It must be emphasized that the bearings in the spindle cartridge during the work described here, were subjected to thrusts that were not in complete agreement with the spindle's design principles. The table movement of this machine was not as accurate or as steady as one would like, but, due to the machine's regular six-monthly maintenance checks, its condition was comparable with the normal standard tool and cutter grinders used in industry today.

Machine B was a more sophisticated device which had been in service for only three months. This machine, a cutter grinder equipped with an automatic indexing head, had a particularly smooth hydraulic table traverse movement; there was no discernible vibration from the spindle drive motor, the hydraulic motor or from the spindle drive belt.

The test grinding was carried out on 24 tungsten carbide specimens using three common cup-wheel types, flaring, straight and dished, each containing diamond abrasive at 100 concentration (4·4 carats/cm³). Other details of the grits used are given in Table 5. Measurement of wheel wear was made to the nearest 0·001 mm under controlled temperature conditions, and after each complete test the wheel was retained in the measuring room for 24 hours to allow it to stabilize before being measured. After grinding, the tungsten carbide specimens were allowed a 20-minute stabilization period before being measured to determine the amount of material removed.

Table 5

WHEEL NO.	WHEEL TYPE	GRIT TYPE	GRIT SIZE
Series I			
1	D11V9	RDA	80/100
2	D11V9	RDA*	80/100
3	D11V9	RDA clad with 10·4% Ni	80/100
4	D11V9	RDA clad with 20·7% Ni	80/100
5	D11V9	RDA clad with 43·3% Ni	80/100
6	D11V9	RDA clad with 53·6% Ni	80/100
Series II			
7	D6A9	RDA clad with 30·5% Ni	120/170
8	D6A9	RDA clad with 30·5% Ni	120/170
9	D6A9	RDA clad with 34·3% Ni	120/170
10	D6A9	RDA clad with 34·3% Ni	120/170
Series III			
11	D12A2	RDA	60/80
12	D12A2	RDA	60/80
13	D12A2	RDA	100/120
14	D12A2	RDA	100/120
15	D12A2	RDA	140/170
16	D12A2	RDA	140/170
17	D12A2	RDA	200/230
18	D12A2	RDA	200/230

* Below specification.

From the measurements obtained, volumetric grinding ratios (amount of TC removed/amount of wheel worn away) were calculated in terms of each machine, and relative efficiencies determined. It soon became clear that the efficiency of machine B was of a much higher level, and one of the most significant points to arise was the considerable reduction in wheel wear on machine B. In fact, the diamond consumption on machine A was found to be 5–6 times greater than that of machine B. The average efficiency figures calculated for the first series of tests produced a factor of 4·7 in favour of machine B; this convincing result was supplemented by a similar figure when the results of the second series of tests were evaluated. The tests in series three, which were carried out with varying infeed, showed that with a smaller infeed (0·025 mm) machine B had an efficiency of 10·5 : 1 compared with machine A. With an infeed of 0·05 mm the performance gap between the two machines widened still further and machine B's relative efficiency increased to 13·6 : 1, thus bringing out the point that machine B was able to cope more efficiently with the larger infeed.

From the results of these tests and from other supporting experience, there is no doubt that the greater cost of the better machine could be more than justified. The price difference of the two machines used in the tests was around £1,750, but it is clear that in many cases, particularly in a tool room, where the tool and cutter grinder is in continual use on one type of job and where the wheel type and infeed remain unchanged, the increased initial outlay can very soon be recovered; after this, of course, very real savings can be obtained. Machine B is in fact a good example of a machine ideally suited for use with the diamond abrasive wheel. It includes all the eight features which the De Beers Diamond Research Laboratory advocates as essential for maximum efficiency when grinding with diamond abrasive wheels:

1. Extremely rigid construction, with no tendency towards "top-heaviness". A wide base on which the machine rests firmly.

2. No vibration transmitted from any of the drive or hydraulic motors into the machine's castings. (If the spindle drive is by belt, there should be no vibration pick-up from the belt.)
3. Coolant stored outside the body casting.
4. Table traverse movement completely free from any "slip-stick" shuddering, particularly at very slow travel speeds.
5. A grinding wheel spindle that runs smoothly and coolly at all speeds. Vibration here is more detrimental to wheel life than any other factor.
6. Minimal movement of the spindle when the grinding load is applied. A cut of 0·025 mm cannot be taken if the spindle can be moved 0·025 mm in its housing.
7. No "give" in the work head–workpiece–tailstock system when the grinding cut is applied.
8. A clamping device that holds the workpiece rigidly. During grinding no vibration of the workpiece can be allowed. Vibration during grinding becomes impact on the face of the wheel and leads to shattering of the diamond in the bond, with consequent increased wheel costs and reduced efficiency. Exactly the same detrimental effect occurs when a wheel is running out of true.

CHATTER VIBRATIONS

In practice, because of poor or inadequate machine design, vibrations often occur which lead to a relative motion between the wheel and the workpiece. As a result, chatter marks are made on the ground surface of the workpiece. The vibrations which cause these chatter marks are called chatter vibrations. Such vibrations lead to (1) a lowering of the surface quality, and to shape and size variations of the workpieces which may lead to their rejection, (2) a shorter tool life because of the necessity for a shorter interval between dressing operations, or greater wear on the grinding wheel, (3) greater wear on the machine, especially on the grinding spindle and bearings, which is not always immediately apparent but ultimately becomes noticeable, and (4) a decrease in and limitation of the machine's capacity, since to maintain stock removal rates would result in an intensification of the chatter vibrations.

Chatter vibrations exert a considerable influence on the cost of grinding. They may be classified as:

(*a*) Natural vibrations
(*b*) Self-excited vibrations
(*c*) Forced vibrations.

(*a*) NATURAL VIBRATIONS

Natural vibrations of the grinding wheel are independent of external influences and can be largely disregarded in relation to machine design.

(*b*) SELF-EXCITED VIBRATIONS

Self-excited vibrations require a supply of energy and the grinding process is itself such a source. A grinding wheel consists of a large number of irregularly bonded abrasive particles with geometrically undefined cutting edges. Around the abrasives the grinding wheel surface needs to be porous to contain the small chips cut off by the abrasive particles during grinding. After a time, blunt areas are worn on the particles, and these worn grits, the number of which continually increases, deform the workpiece material plastically and finally can no longer cut the material. Because of this, the friction forces increase until either the grits are pulled out of the bond or they break, bringing a new sharp cutting edge into contact with the workpiece surface. This is a continual and repetitive process which results in chatter.

In addition to the wearing down of the abrasives, the pores of the grinding wheel can become clogged up. If this occurs, the grinding wheel is lifted from the workpiece. As contact is remade, the grit in contact with the surface pulls a thick chip off the workpiece. This process represents an ideal vibration exciter, which again results in self-excited chatter vibrations.

(*c*) FORCED VIBRATIONS

Forced vibrations are due to external excitation. The unbalance of rotating machine parts and the grinding wheel, defects in bearings, irregularities in the movement of reciprocating masses and hydraulic aggregates, etc., will all result in forced vibrations.

Forced vibrations can be induced by the driving motor, if the rotating magnetic field in the stator deflects the rotor onto an elliptical course. If this happens, two pulsating forces occur per revolution which cause forced vibration at double the mains frequency.

Driving belt defects (for example belt pull), non-endless driving belts, or misaligned bearings also result in forced vibrations, the latter at double the grinding spindle rotational frequency.

WAVINESS

If chatter vibration occurs, its amplitude increases

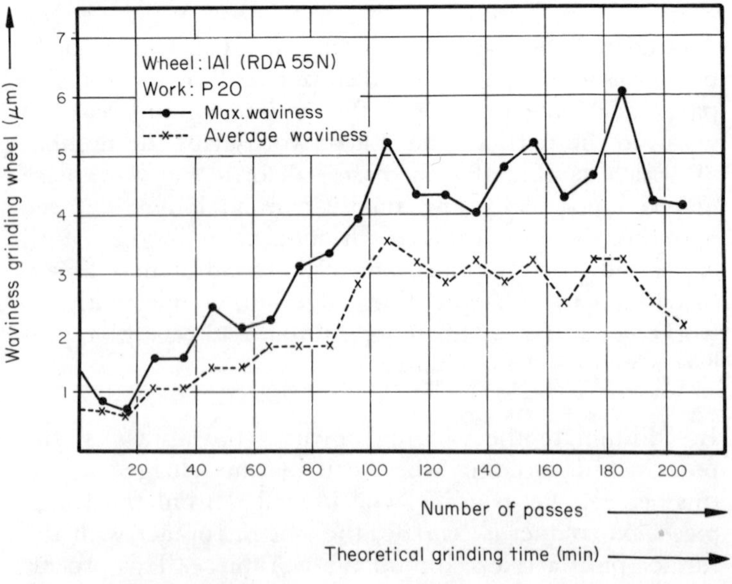

Fig. 15. Depth of waviness as a function of number of grinding passes (grinding time); surface grinding machine.

rapidly as a function of the number of grinding passes, or grinding time. Ultimately the amplitude stabilizes at an undesirable level, dependent on the cutting conditions, the type of wheel and wheel hardness, the workpiece material and machine tool rigidity.

As a consequence of the relative motion between the grinding wheel and the workpiece caused by chatter vibration, a waviness develops in the wheel circumference, the magnitude of which increases rapidly with the grinding time (see Fig. 15). The waviness develops more slowly if the depth of cut is reduced (Fig. 16).

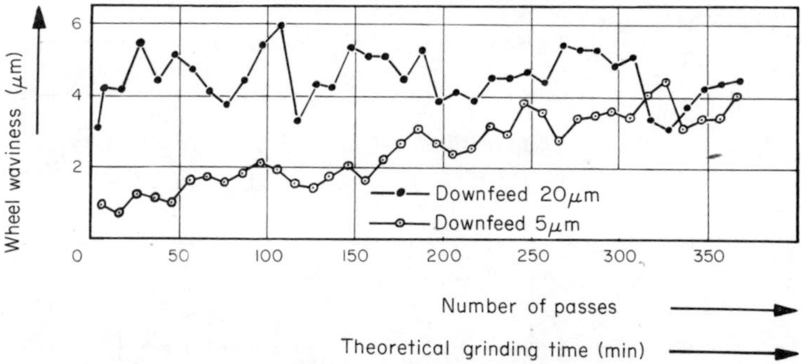

Fig. 16. Effect of downfeed on depth of waviness.

Workpiece quality is affected considerably by chatter vibration and the formation of waviness on the grinding wheel circumference, and deteriorates with increasing vibration intensity.

FREQUENCY DETERMINATION

The determination of the frequency of chatter vibration is important because it shows whether it is caused by a forced or by a self-excited vibration. Once this is known, steps can be taken to avoid or to reduce the vibrations. One method developed by the Diamond Research Laboratory involves recording the vibration amplitudes and from them determining the number of vibrations per time unit (see Figs. 17 and 18). Because chatter vibration often consists of several combined vibrations, analysis by splitting into its components is necessary.

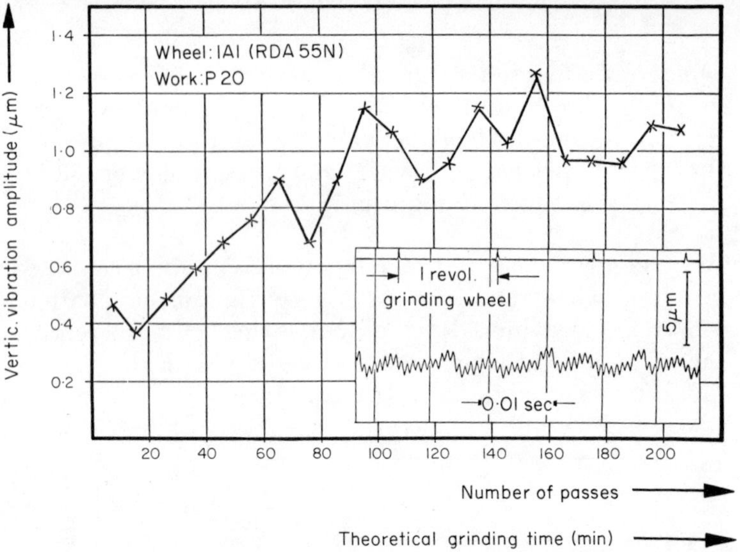

Fig. 17. The chatter vibration amplitude as a function of the number of grinding passes (grinding time); surface grinding machine.

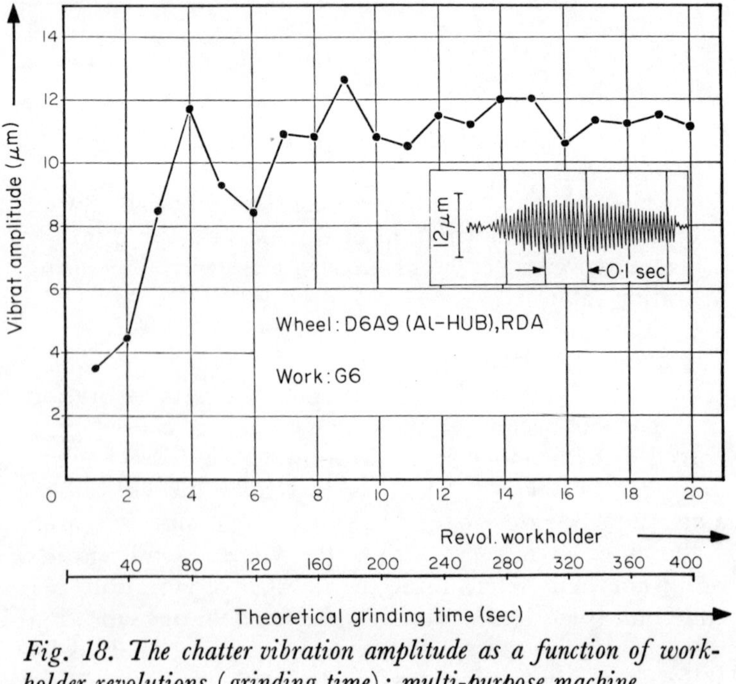

Fig. 18. The chatter vibration amplitude as a function of workholder revolutions (grinding time); multi-purpose machine.

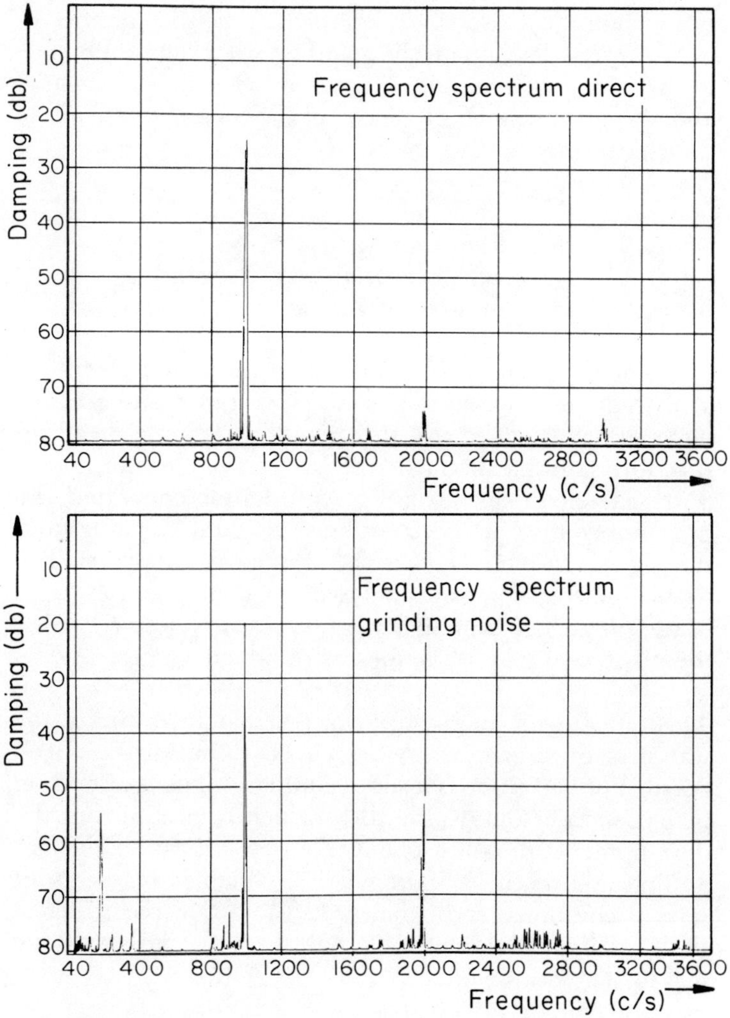

Fig. 19. Frequency spectrum as recorded during grinding; surface grinding machine.

Frequency may also be determined from the number of waves which develop on the grinding wheel circumference in the course of grinding. Knowing the number of waves and the grinding-wheel speed, it is possible to calculate the chatter frequency:

$$F = n.z.$$

where F = frequency (c/s)
n = wheel revolutions per second
z = number of waves

Frequency determination was also investigated by means of a frequency analyser. An acceleration-sensitive transducer was attached to the grinding spindle head to transform the mechanical measured variable into a proportional electrical variable. An adaptor converted the high impedance of the transducer to a lower value, and the signal was relayed to the frequency analyser. The frequency spectrograph analysed frequency compositions in the range 30 c/s to 20 kc/s and measured and recorded the respective amplitude for each frequency (Fig. 19).

Theoretically, it might appear possible to evaluate the waviness of a grinding wheel by ear, but in fact it is possible only if there is a sudden and significant alteration in noise. The sound alteration which occurs during the development of self-excited chatter vibration is not audible above the grinding noise, which is made up of several combined frequencies. An objective recording of the grinding noise is possible only with special electronic measuring equipment.

In the Laboratory, grinding noise was measured by a sensitive microphone which screened out environmental sounds; the signal was recorded and relayed to the frequency analyser. It was found difficult, however, to screen out interfering sounds and measure only the significant chatter noise.

ELIMINATION OF CHATTER VIBRATION

Obviously chatter vibration should be eliminated if

possible, because its existence adds considerably to the cost of grinding. Dynamic balancing of the grinding wheel, pulleys, grinding spindle and electric motor will help to eliminate forced vibration. Quite often, damaged bearings cause vibrations—a large amount of bearing play has the same effect as an unbalanced grinding wheel. The introduction of endless flat belts will also help, and in fact any steps taken to ensure silent running and a high degree of running accuracy of the individual machine parts will be beneficial.

The complete elimination of self-excited chatter vibrations would appear to be impossible. Even changes in machine rigidity and the use of vibration dampers, or alterations in the grinding conditions do not result in the elimination of chatter vibrations. It is, of course, possible to obtain a maximum grinding time before significant chatter vibration occurs.

By decreasing the machine capacity, and thereby reducing the output, the development of the chatter vibration amplitude can be temporarily retarded, but this results in an increase in time cost. Only in cases where a high degree of surface quality in a workpiece is desired is it feasible to reduce the depth of cut in order to retard the development of chatter vibrations. Alteration of the grinding-wheel speed within the optimum working range can also reduce the chatter vibration amplitude, and help avoid the development of regular waves.

Chatter vibrations can also be reduced by means of an artificially induced frequency excitation on the machine tool–grinding wheel–workpiece system. In tests with ultrasonic excitation, a lower chatter vibration amplitude was established, the surface quality of the workpiece was improved, the grinding temperature was reduced and there were no signs of temperature cracks on the ground workpiece surface. In other grinding tests, a counter vibration to the chatter vibration was produced on the machine system in such a way that the relative motion between the grinding wheel and the workpiece disappeared. This method involved complicated technical equipment that would not be feasible in practice. It is not known how far the life expectancy of a machine tool would be diminished by additional ultrasonic vibration excitation.

MODIFYING WHEEL RIGIDITY

A further method of avoiding chatter vibrations, by modifying wheel rigidity, has been investigated by the Diamond Research Laboratory using D6A9 cup wheels. Tests were carried out on tungsten carbide G6 with D6A9 diamond grinding wheels of various rigidities. The

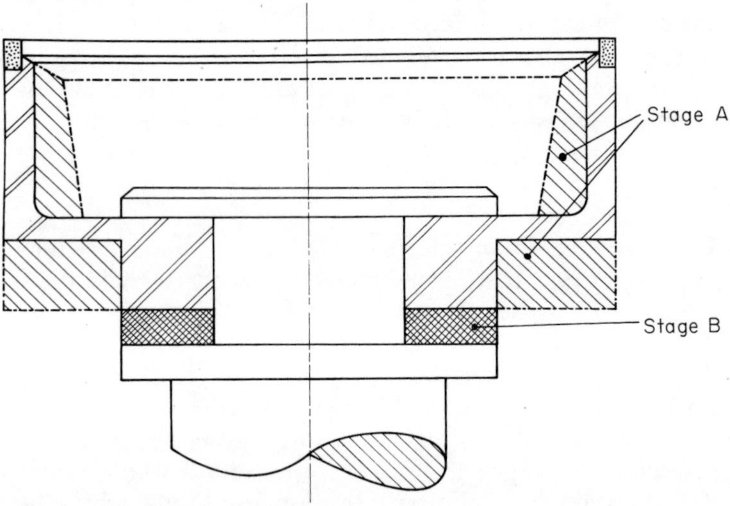

Fig. 20. Cross-section of the test grinding wheel. Stage A: hub material machined off. Stage B: rubber disc only.

wheel flexibility was increased by machining off part of the hub on a lathe (Fig. 20). Improvements of nearly 300% in the grinding ratio (*G*) were obtained. Waviness on the grinding wheel circumference was very much less. The chatter vibration amplitude was low and did not show any increase even after long periods of grinding. The workpiece surface quality was also improved.

In another series of tests, the rigidity of the diamond cup wheels was reduced more simply, without machining the wheel hub, by applying a rubber disc between the wheel and the flange behind the hub. Degrees of flexibility were obtained by choosing rubber discs of varying hardness (e.g. 80, 60, 40 shore).

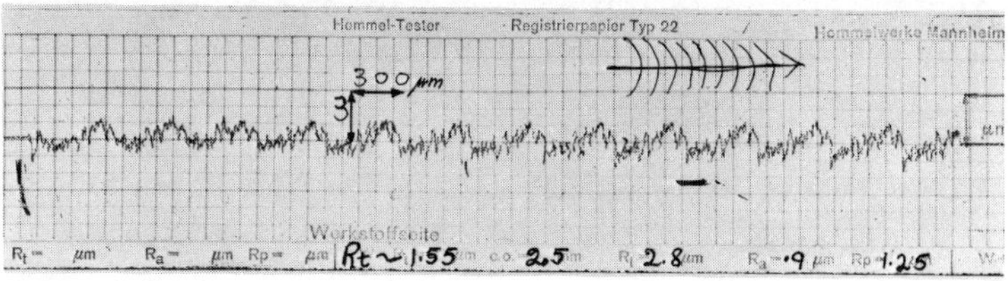

Workpiece surface; grinding wheel: standard arrangement

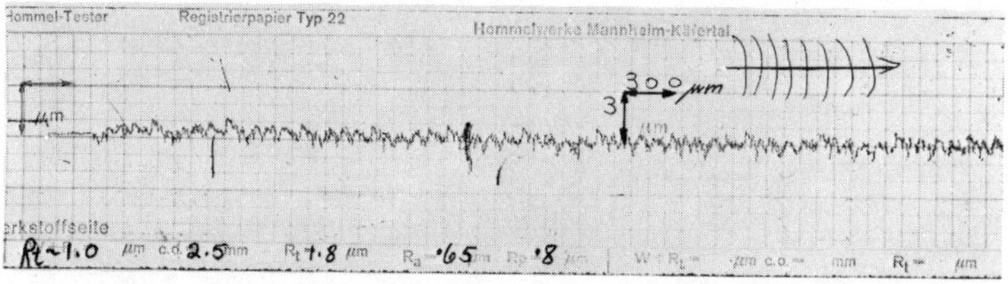

Workpiece surface, grinding wheel: arrangement with rubber disc

Fig. 21. Surface quality of the ground tungsten carbide G6; multi-purpose grinding machine.

Tests on tungsten carbide G6 on a multi-purpose grinding machine gave improvements in the grinding ratio (G) of up to 170%.

Grinding wheel Type 6A9	*Grinding ratio, G*	*Surface finish, R*
Standard arrangement	13·9	1·55 µm
Arrangement with rubber disc	37·6	1·0 µm
Improvement	170%	35%

The workpiece surface was also considerably improved (Fig. 21). Waviness was decreased and showed no regular pattern.

Improvements in grinding performance were obtained

only with the modified diamond cup wheels on a multi-purpose grinding machine. Corresponding tests on a surface grinding machine with peripheral wheels modified to give increased flexibility resulted in decreased grinding quality. It is not clear why a more flexible cup wheel should result in improved performance, whereas a more flexible peripheral wheel gives decreased performance. It would be expected that a more flexible grinding wheel would reduce the coupling stiffness of the machine tool–grinding wheel–workpiece system, giving a lower chatter frequency at increasing vibration amplitude. In general, the coupling stiffness has to be as high as possible to minimize the chatter vibration amplitude. There is obviously much interesting work to be done in this field.

CONCLUSION

Increased machine rigidity and a stiffer workpiece mounting result in longer tool life and better grinding wheel performance.

Many diamond wheels at present in use have to be adapted all too often to an existing machine. Only a machine design which starts basically from the tool—from the diamond wheel itself—and takes into consideration the specific cutting forces and the special physical properties of the diamond will succeed in utilizing to the full all the diamond wheel's potential performance and benefits, i.e. long wheel life, better surface finishes, shorter machining times and low tool and working time costs.

4. Coolant Additives

Because water is such an excellent cooling medium, grinding fluids are often water-based. Many products are available for adding to water to produce grinding fluids, and interesting results have been obtained in comparative tests with them at the DRL when grinding tungsten carbide with resin-bonded diamond wheels.

Although these water-based mixtures are often called coolants, this is really an unsatisfactory name because in

Fig. 22. A tungsten carbide form tool used in the manufacture of components for Concorde is surface ground on a Jones & Shipman 540 machine using a 7 in diameter × ½ in face metal-bond wheel impregnated with De Beers SND-MB. A copious supply of soluble oil coolant is used throughout the operation.

certain circumstances their other functions may be more important. These include (1) lubrication, which tends to increase wheel life by reducing the wear on the diamond crystals and by reducing loading, (2) rust inhibition, which is obviously desirable, and (3) detergent action, which helps keep the wheel sharp and clean. Many grinding liquids also have additives designed to prevent welding of metal chips to each other or to the abrasive.

Fig. 23. A cylindrical component flame-plated with tungsten carbide is ground to a surface finish as fine as 14 micro-in r.m.s. with a resin-bond diamond wheel impregnated with 100 US mesh De Beers SND-RB natural diamond grit. A flood coolant of water +1 to 2% rust inhibitor is used.

When used with resin-bonded diamond wheels, water-based fluids effectively reduce the wheel temperature in the grinding contact area, and therefore help to reduce breakdown of the organic bond, giving increased wheel life. But some additives can have an adverse effect on the wheel-bond material; strong alkaline solutions, for example, can weaken resin bonds. The extent to which a wheel, and therefore its efficiency, may be affected by the grinding fluid, depends to a large extent on the (for the most part) unknown composition of the additives; the DRL decided, therefore, to test the performance of standard resin-bond wheels on tungsten carbide with several different coolants. The experiments showed that increased efficiency levels of the order of 100% can be obtained by taking advantage of certain additives.

The results of these tests, which are of interest to everyone concerned with the use of diamond abrasive grinding wheels, are summarized in Figs. 24 and 25 and Table 6.

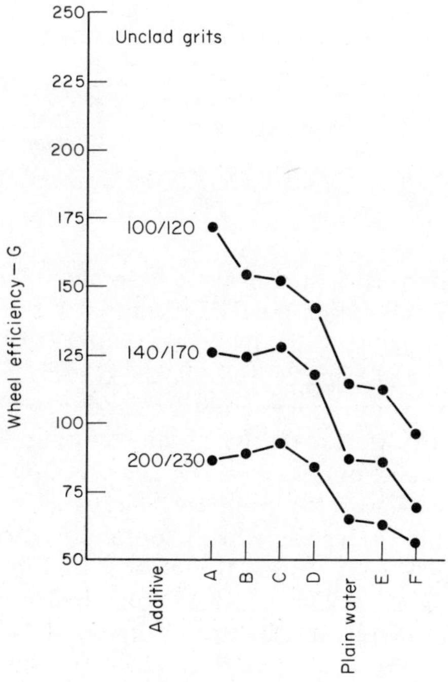

Fig. 24. Effects of different additives on wheel efficiency for unclad grits.

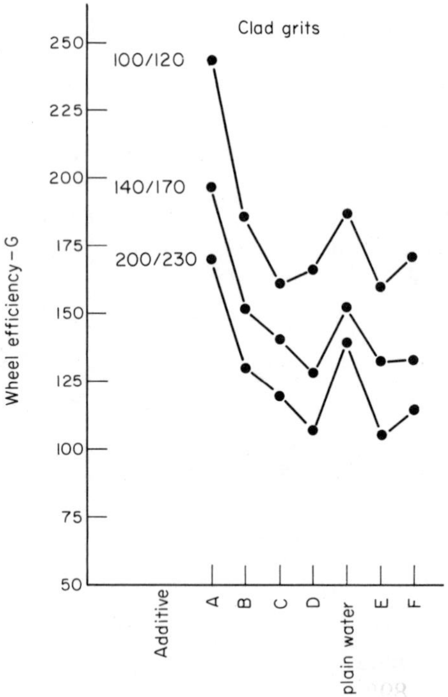

Fig. 25. Effects of different additives on wheel efficiency for clad grits.

The additives used in the tests, designated "A" to "F", are all commercially available standard products, and none is intended specifically for use with diamond wheels. In the graphs of Figs. 24 and 25, the *G* ratio is plotted for each of the proprietary additives and for plain water. Figure 24 shows the results obtained using wheels containing standard unclad grits (De Beers RDA) and Fig. 25 those for wheels containing metal-clad grits (De Beers RDA55N); in each case, wheels containing three different mesh-size grits were tested. It is clear from Fig. 24 that for standard unclad grits, regardless of mesh size, plain water rates quite low in the efficiency range, but in the case of metal-clad grits (Fig. 25) it is close to being the best coolant. It is also clear from a comparison between Figs. 24 and 25 that the relative efficiency levels are considerably higher in the case of the metal-clad grits.

Table 6

GRIT SIZE		100/120		140/170 *G* RATIOS		200/230	
ADDITIVE	GRIT TYPE	ACTUAL	RELA-TIVE	ACTUAL	RELA-TIVE	ACTUAL	RELA-TIVE
A	RDA	170	100	125	100	86	100
	RDA55N	243	143	197	158	170	198
B	RDA	152	100	123	100	88	100
	RDA55N	182	120	148	120	126	143
C	RDA	149	100	124	100	89	100
	RDA55N	158	106	140	113	118	133
D	RDA	141	100	116	100	83	100
	RDA55N	159	113	126	109	104	125
Plain	RDA	114	100	87	100	64	100
water	RDA55N	186	163	152	175	140	219
E	RDA	111	100	85	100	61	100
	RDA55N	160	144	132	155	105	172
F	RDA	90	100	69	100	56	100
	RDA55N	170	189	132	191	113	202

Table 6 shows that in the case of additive "F" the increases in efficiency obtained by changing to the metal-clad grits are 89% for 100/120 mesh, 91% for 140/170 mesh and 102% for 200/230 mesh. Looking at the relative figures for additive "D", however, we see that the increases in efficiency obtained by using a metal-clad grit are only 13% for 100/120 mesh, 9% for 140/170 mesh, and 25% for 200/230 mesh. Undoubtedly, the user would find these increased efficiency levels rather disappointing, although it can be seen from Fig. 24 that this particular additive when used with unclad grits gives an efficiency level higher than additive "F". Again additive "A", which has given the best overall results, enables increased efficiency levels to be obtained from metal-clad grits which, although not as impressive as those obtained using additive "F", are substantial, particularly in the case of the finest mesh size.

The picture which emerges from these results is quite clear. If the user wishes to obtain maximum efficiency from his diamond wheel he must choose the coolant carefully; furthermore, he cannot change from unclad to metal-clad grit without reference to the coolant used and expect automatically to obtain high efficiency increases. If, for example, the user of a 200/230 mesh diamond wheel changed over from an unclad/additive "F" combination to a metal-clad/additive "A" combination he could, on the basis of the Diamond Research Laboratory tests, obtain an efficiency increase of 204%. On the other hand, if he changed from an unclad/additive "C" combination to a metal-clad/additive "D" combination he would obtain an efficiency increase of something less than 17%.

INDEX

M & B MONOGRAPHS

ME/1	**Tribology** E D Hondros	0.263.51588.5
ME/2	**Electron Beam Welding** M J Fletcher	0.263.51707.1
ME/3	**Vacuum Brazing** M J Fletcher	0.263.51708.X
ME/4	**Stress Corrosion Failure** P Greenfield	0.263.51725.X
ME/5	**Glow Discharge Material Processing** R A Dugdale	0.263.51787.X
ME/6	**Diamond Grinding** J Burls	0.263.51791.8
ME/7	**Engineering Applications of Beryllium** P Greenfield	0.263.51838.8
ME/8	**Whiskers** C C Evans	0.263.51844.2
ME/9	**Creep of Metals at High Temperatures** P Greenfield	0.263.51846.9
ME/10	**Cavitation** I S Pearsall	0.263.05000.9
CE/1	**Microbiological Corrosion** G H Booth	0.263.51584.2
CE/2	**Solvent Treatment of Coal** W S Wise	0.263.51706.3
CE/3	**The Fluidised Combustion of Coal** D G Skinner	0.263.51714.4
CE/4	**Carbonisation of Coal** J Gibson D H Gregory	0.263.51774.8
CE/5	**The Principles of Gas Extraction** P F M Paul W S Wise	0.263.51802.7
CE/6	**Metal-Air Batteries** D P Gregory	0.263.51712.8
CE/7	**Fuel Cells** D P Gregory	0.263.51713.6

EE/1	**Threshold Logic** S L Hurst	0.263.51587.7
EE/2	**The Magnetron Oscillator** E Kettlewell	0.263.51641.5
EE/3	**Material for the Gunn Effect** J W Orton	0.263.51773.X
EE/4	**Microcircuit Learning Computers** I Aleksander	0.263.51724.1
EE/5	**DC Conduction in Thin Films** J G Simmons	0.263.51786.1
EE/6	**Reading Machines** J A Weaver	0.263.51839.6
EE/7	**Speech Synthesis** J N Holmes	0.263.51845.0
EE/8	**Glow Discharge Display** G F Weston	0.263.05001.7

Forthcoming titles

Magnesium	P Greenfield
Biotechnology of Industrial Water Conservation	B E Purkiss
Superconductivity	I M Firth